Moon

BY MARION DANE BAUER
ILLUSTRATED BY JOHN WALLACE

Ready-to-Read

Simon Spotlight

New York   London   Toronto   Sydney   New Delhi

For my first grandson, Connor Dane Bauer, with love
—M. D. B.

For Zsa—J. G. W.

SIMON SPOTLIGHT
An imprint of Simon & Schuster Children's Publishing Division
1230 Avenue of the Americas, New York, New York 10020
This Simon Spotlight edition July 2021
For information about special discounts for bulk purchases, please contact
Simon & Schuster Special Sales at
1-866-506-1949 or business@simonandschuster.com.
Manufactured in the United States of America 0521 LAK
2 4 6 8 10 9 7 5 3 1
Library of Congress Cataloging-in-Publication Data
Names: Bauer, Marion Dane, author. | Wallace, John, illustrator.
Title: Moon / by Marion Dane Bauer ; illustrated by John Wallace.
Description: New York, NY : Simon Spotlight, 2021. | Series: Our universe |
Audience: Ages 4–6 | Audience: Grades K–1 | Summary: "Everyone loves
Earth's little neighbor: the Moon! It lights up our night sky, guides our
oceans, and so much more. Beginning readers will love learning all about
the Moon in this new Level 1 Ready-to-Read"—Provided by publisher.
Identifiers: LCCN 2020036045 | ISBN 9781534486423 (paperback) |
ISBN 9781534486430 (hardcover) | ISBN 9781534486447 (ebook)
Subjects: LCSH: Moon—Juvenile literature.
Classification: LCC QB582 .B376 2021 | DDC 523.3—dc23
LC record available at https://lccn.loc.gov/2020036045

Glossary

✦ **asteroids** (say: A-stuh-roidz): small, rocky objects that circle the Sun just as planets do.

✦ **astronauts** (say: A-struh-nots): people who go into outer space. We refer to them as "cosmonauts" when they are from Russia, "spationauts" when they are from France, and "taikonauts" when they are from China.

✦ **eclipse** (say: ih-KLIPS): an event that happens when the Earth gets in between the Sun and the Moon so that sunlight can't reflect off the Moon.

✦ **gravity** (say: GRA-vuh-tee): a pulling force that works across space. The bigger an object is, the more pull it has. Since Earth is larger than the Moon, Earth pulls the Moon into orbit.

✦ **satellite** (say: SA-tuh-lite): an object in space that circles around a bigger object. Earth has one natural satellite, the Moon. There are many man-made satellites that circle Earth too.

Note to readers: Some of these words may have more than one definition. The definitions above match how these words are used in this book.

What is that so big and
bright in our night sky?

Our Moon, of course!

Scientists are not certain
how our Moon came to be.

Many think a large object slammed into our young Earth billions of years ago.

The impact sent rocks and dust flying.

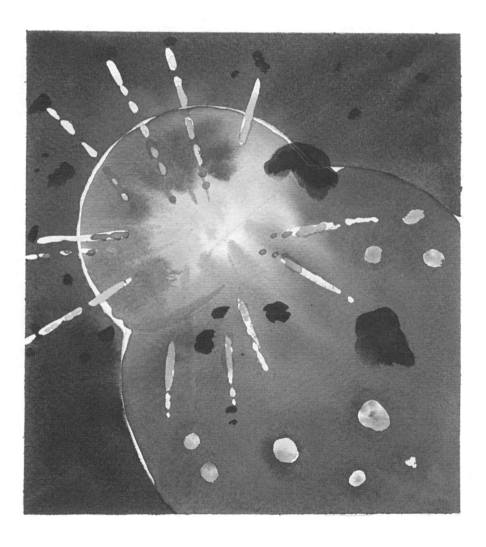

In time the rocks and dust
gathered to form our Moon.

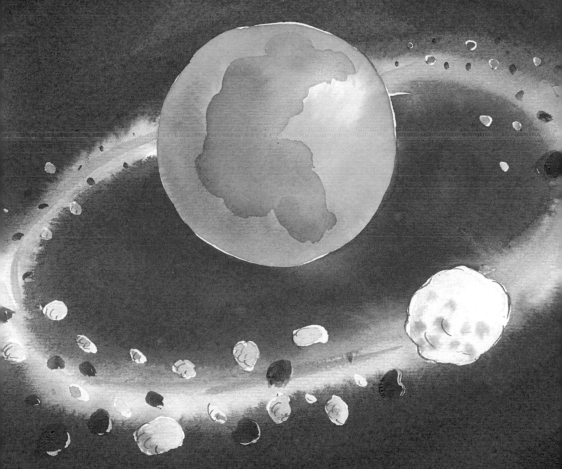

The Moon is Earth's only
natural **satellite**.

It travels around Earth
once every twenty-eight days.

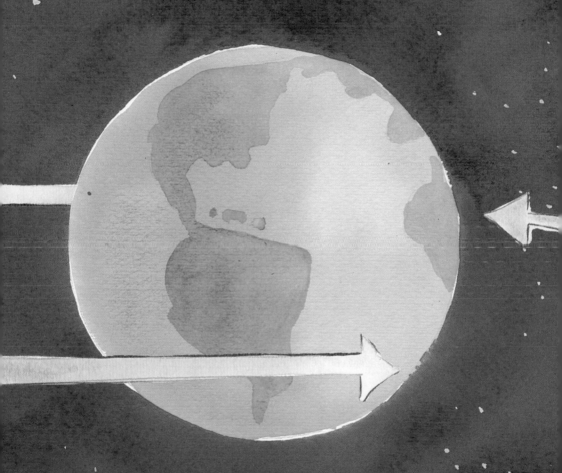

As it travels, the Moon's **gravity** pulls on our oceans.

That pull causes
low and high tides.

Our Moon lights up our night sky, but it has no light of its own.

SUN

The Moon bounces light back
from the Sun.

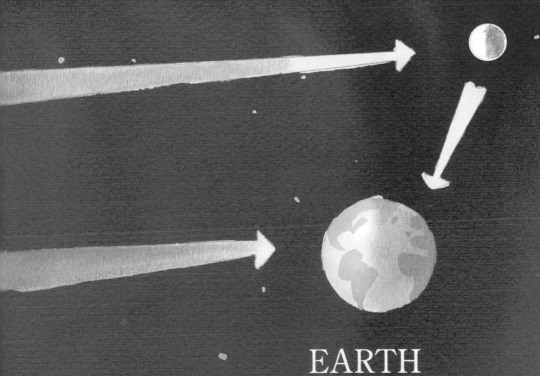

MOON

EARTH

When the Moon passes behind
the Earth, the Sun's light
cannot reach it.

Then we have a
lunar **eclipse**.

We never see the far
side of the Moon.

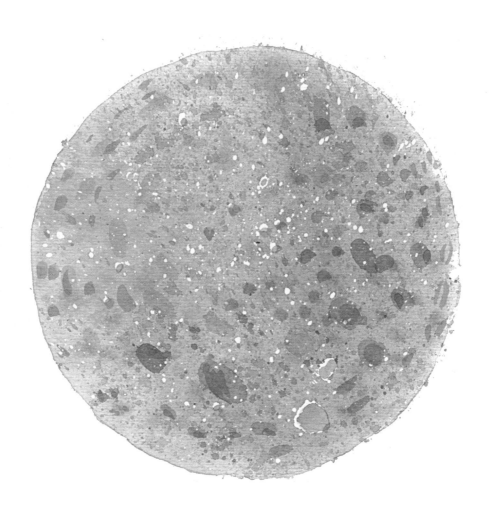

The Moon always keeps the
same face toward Earth.

As the Moon circles
Earth, the Sun lights up
different parts of its face.

1

2

3

4

That is why the Moon's face
keeps changing.

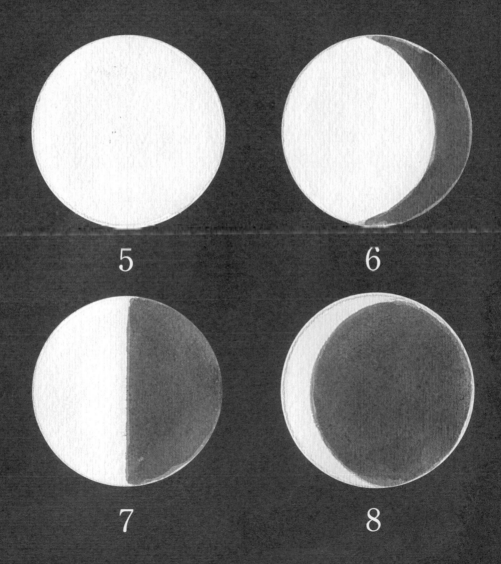

5

6

7

8

Have you ever seen
the man in the Moon?

That face was made long, long ago by **asteroids** striking the Moon.

Some see a woman in the
Moon's craters.

Some see a rabbit

or even a frog!

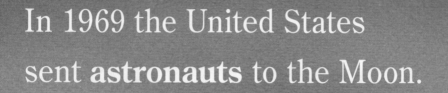

In 1969 the United States
sent **astronauts** to the Moon.

Neil Armstrong became the
first person to ever walk on
the Moon.

The Moon has always been important to humans.

It brightens our nights.
It brightens our lives!

Interesting Facts

✦ Our Moon was formed around four and a half billion years ago.

✦ When the Moon appears to grow fuller, we call it a waxing Moon. When it appears to grow smaller, it is a waning Moon.

✦ The four main Moon phases are new Moon, first quarter Moon, full Moon, and third or last quarter Moon.

✦ There are over 200 moons in our solar system. Many of them orbit Saturn and Jupiter. Mars has two moons. Earth has only one.

✦ In 1959 the Soviets sent a spacecraft on a flyby past the Moon. They even took photos of the far side. That was the first time anyone saw what the far side looked like. Ten years later the United States sent *Apollo 11*, the first manned mission to land on the Moon.

✦ On July 20, 1969, Neil Armstrong stepped onto the Moon's surface and said, "That's one small step for a man, one giant leap for mankind."

✦ People used to wonder what the Moon was made of. Some made funny guesses like the Moon was made of blue cheese! Now that we have been there, we know it is made of rock and metal like our Earth.